The Harper Perennial Resistance Library

In times of crisis, the great works of philosophy help us make sense of the world. The Resistance Library is a special five-book series highlighting short classic works of independent thought that illuminate the nature of TRUTH, humanity's dangerous attraction to AUTHORITARIANISM, the influence of MEDIA and MASS COMMUNICATION, and the philosophy of RESISTANCE—all critical in understanding today's politically charged world.

ON DISOBEDIENCE

ALSO BY ERICH FROMM

The Anatomy of Human Destructiveness

The Art of Loving

Beyond the Chains of Illusion

The Dogma of Christ

Escape from Freedom

The Forgotten Language

Man for Himself

Marx's Concept of Man

May Man Prevail?

Psychoanalysis and Religion

The Sane Society

Sigmund Freud's Mission

Socialist Humanism

You Shall Be as Gods

Zen Buddhism and Psychoanalysis

ON DISOBEDIENCE

Why Freedom Means Saying "NO" to Power

Erich Fromm

The Resistance Library

HARPERPERENNIAL MODERNTHOUGHT

NEW YORK • LONDON • TORONTO • SYDNEY • NEW DELHI • AUCKLAND

HARPERPERENNIAL MODERNTHOUGHT

HarperCollins books may be purchased for educational, business, or sales promotional use. For information please email the Special Markets Department at SPsales@harpercollins.com.

First Harper Perennial Modern Thought edition published 2010.
First Harper Perennial Resistance Library edition published 2019.

Library of Congress Cataloging-in-Publication Data is available upon request.

ISBN 978-0-06-293083-5

HB 01.02.2024

Contents

ON DISOBEDIENCE

I.

Disobedience as a Psychological and Moral Problem

For centuries kings, priests, feudal lords, industrial bosses and parents have insisted that *obedience is a virtue* and that *disobedience is a vice.* In order to introduce another point of view, let us set against this position the following statement: *human history began with an act of disobedience, and it is not unlikely that it will be terminated by an act of obedience.*

Human history was ushered in by an act of disobedience according to the Hebrew and Greek myths. Adam and Eve, living in the Garden of Eden, were part of nature; they were in harmony with it, yet did not transcend it. They were in nature as the fetus is in the womb of the mother. They were human, and at the same time not yet human. All this changed when they disobeyed an order. By breaking the ties with earth and mother, by cutting the umbilical cord, man emerged from a pre-human harmony and was able to take the first step into independence and freedom. The act of disobedience set Adam and Eve free and opened their eyes. They recognized each other as strangers and the world outside them as strange and even hostile. Their act of disobedience broke the primary bond with nature and made them individuals. "Original sin," far from corrupting man, set him free; it was the beginning of history. Man had to leave the Garden of Eden in order to learn to rely on his own powers and to become fully human.

The prophets, in their messianic concept, confirmed the idea that man had been right in disobeying; that he had not been corrupted by his "sin," but freed from the fetters of pre-human harmony. For the prophets, *history* is the place where man becomes human; during its unfolding he develops his powers of reason and of love until he creates a new harmony between himself, his fellow man and nature. This new harmony is described as "the end of days," that period of history in which there is peace between man and man, and between man and nature. It is a "new" paradise created by man himself, and one which he alone could create because he was forced to leave the "old" paradise as a result of his disobedience.

Just as the Hebrew myth of Adam and Eve, so the Greek myth of Prometheus sees all of human civilization based on an act of disobedience. Prometheus, in stealing the fire from the gods, lays the foundation for the evolution of man. There would be no human history were it not for Prometheus' "crime." He, like Adam and Eve, is punished for his disobedience. But he does not repent and ask for forgiveness. On the contrary, he proudly says: "I would rather be chained to this rock than be the obedient servant of the gods."

Man has continued to evolve by acts of disobedience. Not only was his spiritual development possible only because there were men who dared to say no to the powers that be in the name of their conscience or their faith, but also his intellectual development was dependent on the capacity for being disobedient—disobedient to authorities who tried to muzzle new thoughts and to the authority of long-established opinions which declared a change to be nonsense.

If the capacity for disobedience constituted the beginning of human history, obedience might very well, as I have said, cause the end of human history. I am not speaking symbolically or poetically. There is the possibility, or even the probability, that the human race will destroy civilization and even all life upon earth within the next five to ten years. There is no rationality or sense in it. But the fact is that, while we are living technically in the Atomic Age, the majority of men—including most of those who are in power—still live emotionally in the Stone Age; that while our mathematics, astronomy, and the natural sciences are of the twentieth century, most of our ideas about politics, the state, and society lag far behind the age of science. If mankind commits suicide it will be because people will obey those who command them to push the deadly buttons; because they will obey the archaic passions of fear, hate, and greed; because they will obey obsolete clichés of State sovereignty and national honor. The Soviet leaders talk much about revolutions, and we in the "free world" talk much about freedom. Yet they and we discourage disobedience—in the Soviet Union explicitly and by force, in the free world implicitly and by the more subtle methods of persuasion.

But I do not mean to say that all disobedience is a virtue and all obedience a vice. Such a view would ignore the dialectical relationship between obedience and disobedience. Whenever the principles which are obeyed and those which are disobeyed are irreconcilable, an act of obedience to one principle is necessarily an act of disobedience to its counterpart, and vice versa. Antigone is the classic example of this dichotomy. By obeying the inhuman laws of the State, Antigone necessarily would disobey the laws of humanity. By obeying the latter, she must disobey the former. All martyrs of religious faiths, of freedom and of science have had to disobey those who wanted to muzzle them in order to obey their own consciences, the laws of humanity and of reason. If a man can only obey and not disobey, he is a slave; if he can only disobey and not obey, he is a rebel (not a revolutionary); he acts out of anger, disappointment, resentment, yet not in the name of a conviction or a principle.

However, in order to prevent a confusion of terms an important qualification must be made. Obedience to a person, institution or power (heteronomous obedience) is submission; it implies the abdication of my autonomy and the acceptance of a foreign will or judgment in place of my own. Obedience to my own reason or conviction (autonomous obedience) is not an act of submission but one of affirmation. My conviction and my judgment, if authentically mine, are part of me. If I follow them rather than the judgment of others, I am being myself; hence the word *obey* can be applied only in a metaphorical sense and with a meaning which is fundamentally different from the one in the case of "heteronomous obedience."

But this distinction still needs two further qualifications, one with regard to the concept of conscience and the other with regard to the concept of authority.

The word *conscience* is used to express two phenomena which are quite distinct from each other. One is the "authoritarian conscience" which is the internalized voice of an authority whom we are eager to please and afraid of displeasing. This authoritarian conscience is what most people experience when they obey their conscience. It is also the conscience which Freud speaks of, and which he called "Super-Ego." This Super-Ego represents the internalized commands and prohibitions of father, accepted by the son out of fear. Different from the authoritarian conscience is the "humanistic conscience"; this is the voice present in every human being and independent from external sanctions and rewards. Humanistic conscience is based on the fact that as human beings we have an intuitive knowledge of what is human and inhuman, what is conducive of life and what is destructive of life. This conscience serves our functioning as human beings. It is the voice which calls us back to ourselves, to our humanity.

Authoritarian conscience (Super-Ego) is still obedience to a power outside of myself, even though this power has been internalized. Consciously I believe that

I am following *my* conscience; in effect, however, I have swallowed the principles of *power;* just because of the illusion that humanistic conscience and Super-Ego are identical, internalized authority is so much more effective than the authority which is clearly experienced as not being part of me. Obedience to the "authoritarian conscience," like all obedience to outside thoughts and power, tends to debilitate "humanistic conscience," the ability to be and to judge oneself.

The statement, on the other hand, that obedience to another person is *ipso facto* submission needs also to be qualified by distinguishing "irrational" from "rational" authority. An example of rational authority is to be found in the relationship between student and teacher; one of irrational authority in the relationship between slave and master. Both relationships are based on the fact that the authority of the person in command is accepted. Dynamically, however, they are of a different nature. The interests of the teacher and the student, in the ideal case, lie in the same direction. The teacher is satisfied if he succeeds in furthering the student; if he has failed to do so, the failure is his and the student's. The slave owner, on the other hand, wants to exploit the slave as much as possible. The more he gets out of him the more satisfied he is. At the same time, the slave tries to defend as best he can his claims for a minimum of happiness. The interests of slave and master are antagonistic, because what is advantageous to the one is detrimental to the other. The superiority of the one over the other has a different function in each case; in the first it is the condition for the furtherance of the person subjected to the authority, and in the second it is the condition for his exploitation. Another distinction runs parallel to this: rational authority is rational because the authority, whether it is held by a teacher or a captain of a ship giving orders in an emergency, acts in the name of reason which, being universal, I can accept without submitting. Irrational authority has to use force or suggestion, because no one would let himself be exploited if he were free to prevent it.

Why is man so prone to obey and why is it so difficult for him to disobey? As long as I am obedient to the power of the State, the Church, or public opinion, I feel safe and protected. In fact it makes little difference what power it is that I am obedient to. It is always an institution, or men, who use force in one form or another and who fraudulently claim omniscience and omnipotence. My obedience makes me part of the power I worship, and hence I feel strong. I can make no error, since it decides for me; I cannot be alone, because it watches over me; I cannot commit a sin, because it does not let me do so, and even if I do sin, the punishment is only the way of returning to the almighty power.

In order to disobey, one must have the courage to be alone, to err and to sin. But courage is not enough. The capacity for courage depends on a person's state of development. Only if a person has emerged from mother's lap and father's commands, only if he has emerged as a fully developed individual and thus has acquired the capacity to think and feel for himself, only then can he have the courage to say "no" to power, to disobey.

A person can become free through acts of disobedience by learning to say no to power. But not only is the capacity for disobedience the condition for freedom; freedom is also the condition for disobedience. If I am afraid of freedom, I cannot dare to say "no," I cannot have the courage to be disobedient. Indeed, freedom and the capacity for disobedience are inseparable; hence any social, political, and religious system which proclaims freedom, yet stamps out disobedience, cannot speak the truth.

There is another reason why it is so difficult to dare to disobey, to say "no" to power. During most of human history obedience has been identified with virtue and disobedience with sin. The reason is simple: thus far throughout most of history a minority has ruled over the majority. This rule was made necessary by the fact that there was only enough of the good things of life for the few, and only the crumbs remained for the many. If the few wanted to enjoy the good things and, beyond that, to have the many serve them and work for them, one condition was necessary: the many had to learn obedience. To be sure, obedience can be established by sheer force. But this method has many disadvantages. It constitutes a constant threat that one day the many might have the means to overthrow the few by force; furthermore there are many kinds of work which cannot be done properly if nothing but fear is behind the obedience. Hence the obedience which is only rooted in the fear of force must be transformed into one rooted in man's heart. Man must want and even need to obey, instead of only fearing to disobey. If this is to be achieved, power must assume the qualities of the All Good, of the All Wise; it must become All Knowing. If this happens, power can proclaim that disobedience is sin and obedience virtue; and once this has been proclaimed, the many can accept obedience because it is good and detest disobedience because it is bad, rather than to detest themselves for being cowards. From Luther to the nineteenth century one was concerned with overt and explicit authorities.

Luther, the pope, the princes, wanted to uphold it; the middle class, the workers, the philosophers, tried to uproot it. The fight against authority in the State as well as in the family was often the very basis for the development of an independent and daring person. The fight against authority was inseparable from the intellectual mood which characterized the philosophers of the enlightenment and the scientists. This "critical mood" was one of faith in reason, and at the same time of doubt in everything which is said or thought, inasmuch as it is based on tradition, superstition, custom, power. The principles *sapere aude* and *de omnibus est dubitandum*—"dare to be wise" and "of all one must doubt"—were characteristic of the attitude which permitted and furthered the capacity to say "no."

The case of Adolf Eichmann is symbolic of our situation and has a significance far beyond the one which his accusers in the courtroom in Jerusalem were concerned with. Eichmann is a symbol of the organization man, of the alienated bureaucrat for whom men, women and children have become numbers. He is a symbol of all of us. We can see ourselves in Eichmann. But the most frightening thing about him is that after the entire story was told in terms of his own admissions, he was able in perfect good faith to plead his innocence. It is clear that if he were once more in the same situation he would do it again. And so would we—and so do we.

The organization man has lost the capacity to disobey, he is not even aware of the fact that he obeys. At this point in history the capacity to doubt, to criticize and to disobey may be all that stands between a future for mankind and the end of civilization.

II.

Prophets and Priests

It can be said without exaggeration that never was the knowledge of the great ideas produced by the human race as widespread in the world as it is today, and never were these ideas less effective than they are today. The ideas of Plato and Aristotle, of the prophets and of Christ, of Spinoza and Kant, are known to millions among the educated classes in Europe and America. They are taught at thousands of institutions of higher learning, and some of them are preached in the churches of all denominations everywhere. And all this in a world which follows the principles of unrestricted egotism, which breeds hysterical nationalism, and which is preparing for an insane mass slaughter. How can one explain this discrepancy?

Ideas do not influence man deeply when they are only taught as ideas and thoughts. Usually, when presented in such a way, they change other ideas; new thoughts take the place of old thoughts; new words take the place of old words. But all that has happened is a change in concepts and words. Why should it be different? It is exceedingly difficult for a man to be moved by ideas, and to grasp a truth. In order to do that, he needs to overcome deep-seated resistances of inertia, fear of being wrong, or of straying away from the herd. Just to become acquainted with other ideas is not enough, even though these ideas in themselves are right and potent. But ideas do have an effect on man if the idea is lived by the one who teaches it; if it is personified by the teacher, if the idea appears in the flesh. If a man expresses the idea of humility and is humble, then those who listen to him will understand what humility is. They will not only understand, but they will believe that he is talking about a reality, and not just voicing words. The same holds true for all ideas which a man, a philosopher, or a religious teacher may try to convey.

Those who announce ideas—and not necessarily new ones—and at the same time live them we may call *prophets*. The Old Testament prophets did precisely that: they announced the idea that man had to find an answer to his existence, and that this answer was the development of his reason, of his love; and they taught that humility and justice were inseparably connected with love and reason. They lived what they preached. They did not seek power, but avoided it. Not even the power of being a prophet. They were not impressed by might, and they spoke the truth even if this led them to imprisonment, ostracism or death. They were not men who set themselves apart and waited to see what would happen. They responded to their fellow man because they felt responsible. What happened to others happened to them. Humanity was not outside, but within them. Precisely because they saw the truth they felt the responsibility to tell it; they did not threaten, but they showed the *alternatives* with which man was confronted. It is not that a prophet wishes to be a prophet; in fact, only the false ones have the ambition to become prophets. His becoming a prophet is simple enough, because the alternatives which he sees are simple enough. The prophet Amos expressed this idea very succinctly: "The lion has roared, who will not be afraid. God has spoken, who will not be a prophet." The phrase "God has spoken" here means simply that the choice has become unmistakably clear. There can be no more doubt. There can be no more evasion. Hence the man who feels responsible has no choice but to become a prophet, whether he has been herding sheep, tending his vineyards, or developing and teaching ideas. It is the function of the prophet to show reality, to show alternatives and to protest; it is his function to call loudly, to awake man from his customary half-slumber. It is the historical situation which makes prophets, not the wish of some men to be prophets.

Many nations have had their prophets. The Buddha lived his teachings; Christ appeared in the flesh; Socrates died according to his ideas; Spinoza lived them. And they all made a deep imprint on the human race precisely because their idea was manifested in the flesh in each one of them.

Prophets appear only at intervals in the history of humanity. They die and leave their message. The message is accepted by millions, it becomes dear to them. This is precisely the reason why the idea becomes exploitable for others who can make use of the attachment of the people to these ideas, for their own purposes—those of ruling and controlling. Let us call the men who make use of the idea the prophets have announced the *priests*. The prophets live their ideas. The priests administer them to the people who are attached to the idea. The idea has lost its vitality. It has become a formula. The priests declare that it is very important how the idea is formulated; naturally the formulation becomes always important after the experience is dead; how else could one control people by controlling their thoughts, unless there is the "correct" formulation? The priests use the idea to organize men, to control them through controlling the proper expression of the idea, and when they have anesthetized man enough they declare that man is not capable of being awake and of directing his own life, and that they, the priests, act out of duty, or even compassion, when they fulfill the function of directing men who, if left to themselves, are afraid of freedom. It is true not all priests have acted that way, but most of them have, especially those who wielded power.

There are priests not only in religion. There are priests in philosophy and priests in politics. Every philosophical school has its priests. Often they are very learned; it is their business to administer the idea of the original thinker, to impart it, to interpret it, to make it into a museum object and thus to guard it. Then there are the political priests; we have seen enough of them in the last 150 years. They have administered the idea of freedom, to protect the economic interests of their social class. In the twentieth century the priests have taken over the administration of the ideas of socialism. While this idea aimed at the liberation and independence of man, the priests declared in one way or another that man was not capable of being free, or at least that he would not be for a long time. Until then they were obliged to take over, and to decide how the idea was to be formulated, and who was a faithful believer and who was not. The priests usually confuse the people because they claim that they are the successors of the prophet, and that they live what they preach. Yet, while a child could see that they live precisely the opposite of what they teach, the great mass of the people are brainwashed effectively, and eventually they come to believe that if the priests live in splendor they do so as a sacrifice, because they have to represent the great idea; or if they kill ruthlessly they only do so out of revolutionary faith.

No historical situation could be more conducive to the emergence of prophets than ours. The existence of the entire human race is threatened by the madness of preparing nuclear war. Stone-age mentality and blindness have led to the point where the human race seems to be moving rapidly toward the tragic end of its history at the very moment when it is near to its greatest achievement. At this point humanity needs prophets, even though it is doubtful whether their voices will prevail against that of the priests.

Among the few in whom the idea has become manifest in the flesh, and whom the historical situation of mankind has transformed from teachers into prophets, is Bertrand Russell. He happens to be a great thinker, but that is not really essential to his being a prophet. He, together with Einstein and Schweitzer, represents the answer of Western humanity to the threat to its existence, because all three of them have spoken up, have warned, and have pointed out the alternatives. Schweitzer lived the idea of Christianity by working in Lambaréné. Einstein lived the idea of reason and humanism by refusing to join the hysterical voices of nationalism of the German intelligentsia in 1914 and many times after that. Bertrand Russell for many decades expressed his ideas on rationality and humanism in his books; but in recent years he has gone out to the marketplace to show all men that when the laws of the country contradict the laws of humanity, a true man must choose the laws of humanity.

Bertrand Russell has recognized that the idea, even if embodied in one person, gains social significance only if it is embodied in a group. When Abraham argued with God about Sodom's fate, and challenged God's justice, he asked that Sodom be spared if there were only ten just men, but not less. If there were less than ten, that is to say, if there were not even the smallest group in which the idea of justice had become embodied, even Abraham could not expect the city to be saved. Bertrand Russell tries to prove that there are the ten who can save the city. That is why he has organized people, has marched with them, and has sat down with them and been carried off with them in police vans. While his voice is a voice in the wilderness it is, nevertheless, not an isolated voice. It is the leader of a chorus; whether it is the chorus of a Greek tragedy or that of Beethoven's Ninth Symphony only the history of the next few years will reveal.

Among the ideas which Bertrand Russell embodies in his life, perhaps the first one to be mentioned is man's right and duty to disobedience.

By disobedience I do not refer to the disobedience of the "rebel without cause" who disobeys because he has no commitment to life except the one to say "no." This kind of rebellious disobedience is as blind and impotent as its opposite, the conformist obedience which is *incapable* of saying "no." I am speaking of the man who can say "no" because he can affirm, who can disobey precisely because he can obey his conscience and the principles which he has chosen; I am speaking of the revolutionary, not the rebel.

In most social systems obedience is the supreme virtue, disobedience the supreme sin. In fact, in our culture when most people feel "guilty," they are actually feeling afraid because they have been disobedient. They are not really troubled by a moral issue, as they think they are, but by the fact of having disobeyed a command. This is not surprising; after all, Christian teaching has interpreted Adam's disobedience as a deed which corrupted him and his seed so fundamentally that only the special act of God's grace could save man from this corruption. This idea was, of course, in accord with the social function of the Church which supported the power of the rulers by teaching the sinfulness of disobedience. Only those men who took seriously the biblical teachings of humility, brotherliness, and justice rebelled against secular authority, with the result that the Church, more often than not, branded them as rebels and sinners against God. Mainstream Protestantism did not alter this. On the contrary, while the Catholic Church kept alive the awareness of the difference between secular and spiritual authority, Protestantism allied itself with secular power. Luther was only giving the first and drastic expression to this trend when he wrote about the revolutionary German peasants of the sixteenth century, "Therefore let us everyone who can, smite, slay and stab, secretly or openly, remembering that nothing can be more poisonous, hurtful or devilish than a rebel."

In spite of the vanishing of religious terror, authoritarian political systems continued to make obedience the human cornerstone of their existence. The great revolutions in the seventeenth and eighteenth centuries fought against royal authority, but soon man reverted to making a virtue of obedience to the king's successors, whatever name they took. Where is authority today? In the totalitarian countries it is overt authority of the state, supported by the strengthening of respect for authority in the family and in the school. The Western democracies, on the other hand, feel proud at having overcome nineteenth-century authoritarianism. But have they—or has only the character of the authority changed?

This century is the century of the hierarchically organized bureaucracies in government, business, and labor unions. These bureaucracies administer things *and* men as one; they follow certain principles, especially the economic principle of the balance sheet, quantification, maximal efficiency, and profit, and they function essentially as would an electronic computer that has been programmed with these principles. The individual becomes a number, transforms himself into a thing. But just because there is no overt authority, because he is not "forced" to obey, the individual is under the illusion that he acts voluntarily, that he follows only "rational" authority. Who can disobey the "reasonable"? Who can disobey the computer-bureaucracy? Who can disobey when he is not even aware of obeying? In the family and in education the same thing happens. The corruption of the theories of progressive education have led to a method where the child is not told what to do, not given orders, nor punished for failure to execute them. The child just "expresses himself." But, from the first day of his life onward, he is filled with an unholy respect for conformity, with the fear of being "different," with the fright of being away from the rest of the herd. The "organization man" thus reared in the family and in the school and having his education completed in the big organization has opinions, but no convictions; he amuses himself, but is unhappy; he is even willing to sacrifice his life and that of his children in voluntary obedience to impersonal and anonymous powers. He accepts the calculation of deaths which has become so fashionable in the discussions on thermonuclear war: half the population of a country dead—"quite acceptable"; two-thirds dead—"maybe not."

The question of disobedience is of vital importance today. While, according to the Bible, human history began with an act of disobedience—Adam and Eve—while, according to Greek myth, civilization began with Prometheus' act of disobedience, it is not unlikely that human history will be terminated by an act of obedience, by the obedience to authorities who themselves are obedient to archaic fetishes of "State sovereignty," "national honor," "military victory," and who will give the orders to push the fatal buttons to those who are obedient to them and to their fetishes.

Disobedience, then, in the sense in which we use it here, is an act of the affirmation of reason and will. It is not primarily an attitude directed *against* something, but *for* something: for man's capacity to see, to say what he sees, and to refuse to say what he does not see. To do so he does not need to be aggressive or rebellious; he needs to have his eyes open, to be fully awake, and willing to take the responsibility to open the eyes of those who are in danger of perishing because they are half asleep.

Karl Marx once wrote that Prometheus, who said that he "would rather be chained to his rock than to be the obedient servant of the gods," is the patron saint of all philosophers. This consists in renewing the Promethean function of life itself. Marx's statement points very clearly to the problem of the connection between philosophy and disobedience. Most philosophers were not disobedient to the authorities of their time. Socrates obeyed by dying, Spinoza declined the position of a professor rather than to find himself in conflict with authority, Kant was a loyal citizen, Hegel exchanged his youthful revolutionary sympathies for the glorification of the State in his later years. Yet, in spite of this, Prometheus was their patron saint. It is true, they remained in their lecture halls and their studies and did not go to the marketplace, and there were many reasons for this which I shall not discuss now. But as philosophers they were disobedient to the authority of traditional thoughts and concepts, to the clichés which were believed and taught. They were bringing light to darkness, they were waking up those who were half asleep, they "dared to know."

The philosopher is disobedient to clichés and to public opinion because he is obedient to reason and to mankind. It is precisely because reason is universal and transcends all national borders, that the philosopher who follows reason is a citizen of the world; man is his object—not this or that person, this or that nation. The world is his country, not the place where he was born.

Nobody has expressed the revolutionary nature of thought more brilliantly than Bertrand Russell. In *Principles of Social Reconstruction* (1916), he wrote:

> *Men fear thought more than they fear anything else on earth—more than ruin, more even than death. Thought is subversive and revolutionary, destructive and terrible;*

thought is merciless to privilege, established institutions, and comfortable habits; thought is anarchic and lawless, indifferent to authority, careless of the well-tried wisdom of the ages. Thought looks into the pit of hell and is not afraid. It sees man, a feeble speck, surrounded by unfathomable depths of silence; yet bears itself proudly, as unmoved as if it were lord of the universe. Thought is great and swift and free, the light of the world, and the chief glory of man.

But if thought is to become the possession of many, not the privilege of the few, we must have done with fear. It is fear that holds men back—fear lest their cherished beliefs should prove delusions, fear lest the institutions by which they live should prove harmful, fear lest they themselves should prove less worthy of respect than they have supposed themselves to be. "Should the working man think freely about property? Then what will become of us, the rich? Should young men and young women think freely about sex? Then what will become of morality? Should soldiers think freely about war? Then what will become of military discipline? Away with thought! Back into the shades of prejudice, lest property, morals, and war should be endangered! Better men should be stupid, slothful, and oppressive than that their thoughts should be free. For if their thoughts were free they might not think as we do. And at all costs this disaster must be averted." So the opponents of thought argue in the unconscious depths of their souls. And so they act in their churches, their schools, and their universities.

Bertrand Russell's capacity to disobey is rooted, not in some abstract principle, but in the most real experience there is—in the love of life. This love of life shines through his writings as well as through the person. It is a rare quality today, and especially rare in the very countries where men live in the midst of plenty. Many confuse thrill with joy, excitement with interest, consuming with being. The necrophilous slogan "Long live death," while consciously used only by the fascists, fills the hearts of many people living in the lands of plenty, although they are not aware of it themselves. It seems that in this fact lies one of the reasons which explain why the majority of people are resigned to accept nuclear war and the ensuing destruction of civilization and take so few steps to prevent this catastrophe. Bertrand Russell, on the contrary, fights against the threatening slaughter, not because he is a pacifist or because some abstract principle is involved, but precisely because he is a man who loves life.

For the very same reason he has no use for those voices which love to harp on the evilness of man, in fact thus saying more about themselves and their own gloomy moods than about men. Not that Bertrand Russell is a sentimental romantic. He is a hard-headed, critical, caustic realist; he is aware of the depth of evil and stupidity to be found in the heart of man, but he does not confuse this fact with an alleged innate corruption which serves to rationalize the outlook of those who are too gloomy to believe in man's gift to create a world in which he can feel himself to be at home. "Except for those rare spirits," wrote Russell in *Mysticism and Logic: A Free Man's Worship* (1903), "that are born without sin, there is a cavern of darkness to be traversed before that temple can be entered. The gate of the cavern is despair, and its floor is paved with the gravestones of abandoned hopes. There Self must die; there the eagerness, the greed of untamed desire must be slain, for only so can the soul be freed from the empire of Fate. But out of the cavern the Gate of Renunciation leads again to the daylight of wisdom, by whose radiance a new insight, a new joy, a new tenderness, shine forth to gladden the pilgrim's heart." And later, in *Philosophical Essays* (1910), he wrote: "But for those who feel that life on this planet would be a life in prison if it were not for the windows onto a greater world beyond; for those to whom a belief in man's omnipotence seems arrogant, who desire rather the Stoic freedom that comes of mastery over the passions than the Napoleonic domination that sees the kingdoms of this world at its feet—in a word, to men who do not find Man an adequate object of their worship, the pragmatist's world will seem narrow and petty, robbing life of all that gives it value, and making Man himself smaller by depriving the universe which he contemplates of all its splendour." His views on the alleged evilness of man, Russell expressed brilliantly in the *Unpopular Essays* (1950): "Children, after being limbs of Satan in traditional theology and mystically illuminated angels in the minds of educational reformers, have reverted to being little devils—not theological demons inspired by the Evil One, but scientific Freudian abominations inspired by the Unconscious. They are, it must be said, far more wicked than they were in the diatribes of the monks; they display, in modern textbooks, an ingenuity and persistence in sinful imaginings to which in the past there was nothing comparable except St. Anthony. Is all this the objective truth at last? Or is it merely an adult imaginative compensation for being no longer allowed to wallop the little pests? Let the Freudians answer, each for the others." One more quotation from Russell's writings which shows how deeply this humanist thinker has experienced this joy of living. "The lover," he wrote in *The Scientific Outlook* (1931), "the poet, and the mystic find a fuller satisfaction than the seeker after power can ever know, since they can retain the object of their love, whereas the seeker after power must be perpetually engaged in some fresh manipulation if he is not to suffer from a sense of emptiness. When I come to die I shall not feel I have lived in vain. I have seen the earth turn red at evening, the dew sparkling in the morning, and the snow shining under a frosty sun; I have smelt rain after drought, and have heard the stormy Atlantic beat upon the granite shores of Cornwall. Science may bestow these and other joys among more people than could otherwise enjoy them. If so, its power will be wisely used. But when it takes out of life the moments to which life owes its values, science will not deserve admiration, however cleverly and however elaborately it may lead men along the road to despair."

Bertrand Russell is a scholar, a man who believes in reason. But how different he is from the many men whose profession is the same: scholarship. With these the thing that counts is the intellectual grasp of the world. They feel certain that their intellect exhausts reality, and that there is nothing of significance which cannot be grasped by it. They are skeptical toward everything which cannot be caught in an intellectual formula, but they are naively unskeptical toward their own scientific approach. They are more interested in the results of their thoughts than in the process of enlightenment which occurs in the inquiring person. Russell spoke of this kind of intellectual procedure when discussing pragmatism in his *Philosophical Essays* (1910): "Pragmatism," he wrote, "appeals to the temper of mind which finds on the surface of this planet the whole of its imaginative material; which feels confident of progress, and unaware of nonhuman limitations to human power; which loves battle, with all the attendant risks, because it has no real doubt that it will achieve victory; which desires religion, as it desires railways and electric light, as a comfort and a help in the affairs of this world, not as providing nonhuman objects to satisfy the hunger for perfection and for something to be worshipped without reserve."

For Russell, in contrast to the pragmatist, rational thought is not the quest for certainty, but an adventure, an act of self-liberation and of courage, which changes the thinker by making him more awake and more alive.

Bertrand Russell is a man of faith. Not of faith in the theological sense, but of faith in the power of reason, faith in man's capacity to create his own paradise through his own efforts. "As geological time is reckoned," so he wrote in *Man's Peril from the Hydrogen Bomb* (1954), "Man has so far existed only for a very short period—

1,000,000 years at the most. What he has achieved, especially during the last 6,000 years, is something utterly new in the history of the Cosmos, so far at least as we are acquainted with it. For countless ages the sun rose and set, the moon waxed and waned, the stars shone in the night, but it was only with the coming of Man that these things were understood. In the great world of astronomy and in the little world of the atom, Man has unveiled secrets which might have been thought undiscoverable. In art and literature and religion, some men have shown a sublimity of feeling which makes the species worth preserving. Is all this to end in trivial horror because so few are able to think of Man rather than of this or that group of men? Is our race so destitute of wisdom, so incapable of impartial love, so blind even to the simplest dictates of self-preservation, that the last proof of its silly cleverness is to be the extermination of all life on our planet?—for it will be not only men who will perish, but also the animals and plants, whom no one can accuse of communism or anticommunism.

"I cannot believe that this is to be the end. I would have men forget their quarrels for a moment and reflect that, if they will allow themselves to survive, there is every reason to expect the triumphs of the future to exceed immeasurably the triumphs of the past. There lies before us, if we choose, continual progress in happiness, knowledge, and wisdom. Shall we, instead, choose death, because we cannot forget our quarrels? I appeal, as a human being to human beings: remember your humanity, and forget the rest. If you can do so, the way lies open to a new Paradise; if you cannot, nothing lies before you but universal death."

This faith is rooted in a quality without which neither his philosophy nor his fight against war could be understood: *his love for life.*

To many people this may not mean much; they believe that everybody loves life. Does he not cling to it when it is threatened, does he not have a great deal of fun in life and plenty of thrilling excitement?

In the first place, people do not cling to life when it is threatened; how else could one explain their passivity before the threat of nuclear slaughter? Furthermore, people confuse excitement with joy, thrill with love of life. They are "without joy in the midst of plenty." The fact is that all the virtues for which capitalism is praised—individual initiative, the readiness to take risks, independence—have long disappeared from industrial society and are to be found mainly in westerns and among gangsters. In bureaucratized, centralized industrialism, regardless of political ideology, there is an increasing number of people who are fed up with life and willing to die in order to get over their boredom. They are the ones who say "better dead than red," but deep down their motto is "better dead than alive." As I mentioned earlier, the extreme form of such an orientation was to be found among those fascists whose motto was "Long live death." Nobody recognized this more clearly than did Miguel de Unamuno when he spoke for the last time in his life at the University of Salamanca, where he was Rector at the time of the beginning of the Spanish Civil War; the occasion was a speech by General Millán Astray, whose favorite motto was "Viva la Muerte!" (Long live death!) and one of his followers shouted it from the back of the hall. When the general had finished his speech Unamuno rose and said: ". . . Just now I heard a necrophilous and senseless cry: 'Long live death!' And I, who have spent my life shaping paradoxes which have aroused the uncomprehending anger of others, I must tell you, as an expert authority, that this outlandish paradox is repellent to me. General Millán Astray is a cripple. Let it be said without any slighting undertone. He is a war invalid. So was Cervantes. Unfortunately there are too many cripples in Spain just now. And soon there will be even more of them if God does not come to our aid. It pains me to think that General Millán Astray should dictate the pattern of mass psychology. A cripple who lacks the spiritual greatness of a Cervantes is wont to seek ominous relief in causing mutilation around him." At this Millán Astray was unable to restrain himself any longer. *"Abajo la inteligencia!"* (Down with intelligence!) he shouted. "Long live death!" There was a clamor of support for this remark from the Falangists. But Unamuno went on: "This is the temple of the intellect. And I am its high priest. It is you who profane its sacred precincts. You will win, because you have more than enough brute force. But you will not convince. For to convince you need to persuade. And in order to persuade you would need what you lack: Reason and Right in the struggle. I consider it futile to exhort you to think of Spain. I have done."

However, the attraction to death which Unamuno called necrophilia is not a product of fascist thought alone. It is a phenomenon deeply rooted in a culture which is increasingly dominated by the bureaucratic organizations of the big corporations, governments, and armies, and by the central role of man-made things, gadgets, and machines. This bureaucratic industrialism tends to transform human beings into things. It tends to replace nature by technical devices, the organic by the inorganic.

One of the earliest expressions of this love for destruction and for machines, and of the contempt for woman (woman is a manifestation of life for man just as man is a manifestation of life for woman), is to be found in the futuristic manifesto (by Marinetti in 1909), one of the intellectual forerunners of Italian fascism. He wrote:

> *. . . 4. We declare that the world's splendor has been enriched by a new beauty; the beauty of speed. A racing motor-car, its frame adorned with great pipes, like snakes with explosive breath . . . a roaring motor-car, which looks as though running on a shrapnel is more beautiful than the Victory of Samothrace.*
>
> *5. We shall sing of the man at the steering wheel, whose ideal stem transfixes the Earth, rushing over the circuit of her orbit.*
>
> *. . . 8. Why should we look behind us, when we have to break in the mysterious portals of the Impossible? Time and Space died yesterday. Already we live in the absolute, since we have already created speed, eternal and ever-present.*
>
> *9. We wish to glorify War—the only health-giver of the world—militarism, patriotism, the destructive arm of the Anarchist, the beautiful Ideas that kill, the contempt for woman.*
>
> *10. We wish to destroy the museums, the libraries, to fight against moralism, feminism, and all opportunistic and utilitarian meannesses.*

There is indeed no greater distinction among human beings than that between those who love life and those who love death. This love of death is a typically human acquisition. Man is the only animal that can be bored, he is the only animal that can love death. While the impotent man (I am not referring to sexual impotence) cannot create life, he can destroy it and thus transcend it. The love of death in the midst of living is the ultimate perversion. There are some who are the true necrophiles—and they salute war and promote it, even though they are mostly not aware of their motivation and rationalize their desires as serving life, honor, or freedom. They are probably the minority; but there are many who have never made the choice between life and death, and who have escaped in busy-ness in order to hide this. They do not salute destruction, but they also do not salute life. They lack the joy of life which would be necessary to oppose war vigorously.

Goethe once said that the most profound distinction between various historical periods is that between belief and disbelief, and added that all epochs in which belief dominates are brilliant, uplifting, and fruitful, while those in which disbelief dominates vanish because nobody cares to devote himself to the unfruitful. The "belief" Goethe spoke of is deeply rooted in the love of life. Cultures which create the conditions for loving life are also cultures of belief; those which cannot create this love also cannot create belief.

Bertrand Russell is a man of belief. In reading his books and in watching his activities for peace his love of life seems to me the mainspring of his whole person. He warns the world of impending doom precisely as the prophets did, because he loves life and all its forms and manifestations. He, again like the prophets, is not a determinist who claims that the historical future is already determined; he is an "alternativist" who sees that what is determined are certain limited and ascertainable alternatives. Our alternative is that between the end of the nuclear arms race—and destruction. Whether the voice of this prophet will win over the voices of doom and weariness depends on the degree of vitality the world and especially the younger generation has preserved. If we are to perish we cannot claim not to have been warned.

III.

Let Man Prevail

When the medieval world was torn wide open, Western man seemed to be headed for the final fulfillment of his keenest dreams and visions. He freed himself from the authority of a totalitarian Church, the weight of traditional thought, the geographical limitations of a half-discovered globe. He discovered nature and the individual. He became aware of his own strength, of his capacity to make himself the ruler over nature and over traditionally given circumstances. He believed that he would be capable of achieving a synthesis between his newborn sense of strength and rationality and the spiritual values of his humanistic-spiritual tradition, between the prophetic idea of the messianic time of peace and justice to be achieved by mankind in the historical process and the Greek tradition of theoretical thought. In the centuries following the Renaissance and the Reformation, he built a new science which eventually led to the release of hitherto unheard-of productive powers and to the complete transformation of the material world. He created political systems which seem to guarantee the free and productive development of the individual; he reduced the time of work to such an extent that Western man is free to enjoy hours of leisure to an extent his forefathers had hardly dreamed of.

Yet where are we today?

The world is divided into two camps, the capitalist and the communist camp. Both camps believe that they have the key to the fulfillment of the human hopes of generations past; both maintain that, while they must coexist, their systems are incompatible.

Are they right? Are they not both in the process of converging into a new industrial neo-feudalism, into industrial societies, led and manipulated by big, powerful bureaucracies—societies in which the individual becomes a well-fed and well-entertained automaton who loses his individuality, his independence and his humanity? Have we to resign ourselves to the fact that we can master nature and produce goods in an ever-increasing degree, but that we must give up the hope for a new world of solidarity and justice; that this ideal will be lost in an empty technological concept of "progress"?

Is there no other alternative than that between capitalist and communist managerial industrialism? Can we not build an industrial society in which the individual retains his role as an active, responsible member who controls circumstances, rather than being controlled by them? Are economic wealth and human fulfillment really incompatible?

These two camps not only compete economically and politically, they are both set against each other in deadly fear of an atomic attack which will wipe out both, if not all civilization. Indeed, man has created the atomic bomb; it is the result of one of his greatest intellectual achievements. But he has lost the mastery over his own creation. The bomb has become his master, the forces of his own creation have become his most dangerous enemy.

Is there still time to reverse this course? Can we succeed in changing it and becoming the masters of circumstances, rather than allowing circumstances to rule us? Can we overcome the deep-seated roots of barbarism which make us try to solve problems in the only way in which they can *never* be solved—by force, violence, and killing? Can we close the gap between our great intellectual achievement and our emotional and moral backwardness?

In order to answer these questions, a more detailed examination of Western man's present position is necessary.

To most Americans the case for the success of our mode of industrial organization seems to be clear and overwhelming. New productive forces—steam, electricity, oil, and atomic energy—and new forms of organization of work—central planning, bureaucratization, increased division of labor, automation—have created a material wealth in the most advanced industrial countries which has done away with the extreme poverty in which the majority of their populations lived a hundred years ago.

Working hours have been reduced from seventy to forty hours per week in the last hundred years, and with increasing automation an ever-shorter working day may give man an undreamed-of amount of leisure. Basic education has been brought to every child; higher education to a considerable percentage of the total population. Movies, radio, television, sports, and hobbies fill out the many hours which man now has for his leisure.

Indeed, it seems that for the first time in history the vast majority—and soon all men—in the Western world will be primarily concerned with living, rather than with the struggle to secure the material conditions for living. It seems that the fondest dreams of our forefathers are close to their realization, and that the Western world has found the answer to the question what the "good life" is.

While the majority of men in North America and Western Europe still share this outlook, there are an increasing number of thoughtful and sensitive persons who see the flaws in this enticing picture. They notice, first of all, that even within the richest country in the world, the U.S.A., about one-fifth of the population does not share in the good life of the majority, that a considerable number of our fellow citizens have not reached the material standard of living which is the basis for a dignified human existence. They are aware, furthermore, that more than two-thirds of the human race, those who for centuries were the object of Western colonialism, have a standard of living from ten to twenty times lower than ours, and have a life expectancy half that of the average American.

They are struck by the irrational contradictions which beset our system. While there are millions in our own midst, and hundreds of millions abroad, who do not have enough to eat, we restrict agricultural production and, in addition, spend hundreds of millions each year in storing our surplus. We have affluence, but we do not have amenity. We are wealthier, but we have less freedom. We consume more, but we are emptier. We have more atomic weapons, but we are more defenseless. We have more education, but we have less critical judgment and convictions. We have more religion, but we become more materialistic. We speak of the American tradition which, in fact, is the spiritual tradition of radical humanism, and we call "un-American" those who want to apply the tradition to present-day society.

However, even if we comfort ourselves, as many do, with the assumption that it is only a matter of a few generations until the West and eventually the whole world will have reached economic affluence, the question arises: *What has become of man and where is he going if we continue on the road our industrial system has taken?*

In order to understand how those elements by which our system succeeded in solving some of its *economic* problems are leading to an increasing failure to solve the *human* problem, it is necessary to examine the features which are characteristic of twentieth-century capitalism.

Concentration of capital led to the formation of giant enterprises, managed by hierarchically organized bureaucracies. Large agglomerations of workers work together, part of a vast organized production machine which, in order to run at all, must run smoothly, without friction, without interruption. The individual worker and clerk become a cog in this machine; their function and activities are determined by the whole structure of the organization in which they work. In the large enterprises, legal ownership of the means of production has become separated from the management and has lost importance. The big enterprises are run by bureaucratic management, which does not own the enterprise legally, but socially. These managers do not have the qualities of the old owner—individual initiative, daring, risk-taking—but the qualities of the bureaucrat—lack of individuality, impersonality, caution, lack of imagination. They administer things and persons, and relate to persons as to things. This managerial class, while it does not own the enterprise legally, controls it factually; it is responsible, in an effective way, neither to the stockholders nor to those who work in the enterprise. In fact, while the most important fields of production are in the hands of the large corporations, these corporations are practically ruled by their top employees. The giant corporations which control the economic, and to a large degree the political, destiny of the country constitute the very opposite of the democratic process; *they represent power without control by those submitted to it.*

Aside from the industrial bureaucracy, the vast majority of the population is administered by still other bureaucracies. First of all, by the governmental bureaucracy (including that of the armed forces) which influences and directs the lives of many millions in one form or another. More and more the industrial, military and governmental bureaucracies are becoming intertwined, both in their activities and, increasingly, in their personnel. With the development of ever greater enterprises, unions have also developed into big bureaucratic machines in which the individual member has very little to say. Many union chiefs are managerial bureaucrats, just as industrial chiefs are.

All these bureaucracies have no plan, and no vision; and due to the very nature of bureaucratic administration, this has to be so. When man is transformed into a thing and managed like a thing, his managers themselves become things; and things have no will, no vision, no plan.

With the bureaucratic management of people, the democratic process becomes transformed into a ritual. Whether it is a stockholders' meeting of a big enterprise, a political election or a union meeting, the individual has lost almost all influence to determine decisions and to participate actively in the making of decisions. Especially in the political sphere, elections become more and more reduced to plebiscites in which the individual can express preference for one of two slates of professional politicians, and the best that can be said is that he is governed with his consent. But the means to bring about this consent are those of suggestion and manipulation and, with all that, the most fundamental decisions—those of foreign policy which involve peace and war—are made by small groups which the average citizen hardly even knows.

The political ideas of democracy, as the founding fathers of the United States conceived them, were not purely political ideas. They were rooted in the spiritual tradition which came to us from prophetic Messianism, the gospels, humanism, and from the enlightenment philosophers of the eighteenth century. All these ideas and movements were centered around one hope: that man, in the course of his history, can liberate himself from poverty, ignorance and injustice, and that he can build a society of harmony, peace and union between man and man and between man and nature. The idea that history has a goal and the faith in man's perfectability within the historical process have been the most specific elements of Occidental thought. They are the soil in which the American tradition is rooted and from which it draws its strength and vitality. What has happened to the idea of the perfectability of man and of society? It has deteriorated into a flat concept of "progress," into a vision of the production of more and better *things*, rather than standing for the birth of the fully alive and productive *man*. Our political concepts have today lost their spiritual roots. They have become matters of expediency, judged by the criterion of whether they help us to a higher standard of living and to a more effective form of political administration. Having lost their roots in the hearts and longings of man, they have become empty shells, to be thrown away if expediency warrants it.

The individual is managed and manipulated not only in the sphere of production, but also in the sphere of consumption, which allegedly is the one in which the individual expresses his free choice. Whether it is the consumption of food, clothing, liquor, cigarettes, movies, or television programs, a powerful suggestion apparatus is employed with two purposes: first, to constantly increase the individual's appetite for new commodities, and second, to direct these appetites into the channels most profitable for industry. The very size of the capital investment in the consumer-goods industry and the competition between a few giant enterprises make it necessary not to leave consumption to chance, nor to leave the consumer a free choice of whether he wants to buy more and what he wants to buy. His appetites have to be constantly whetted, tastes have to be manipulated, managed, and made predictable. Man is transformed into the "consumer," the eternal suckling, whose one wish is to consume more and "better" things.

While our economic system has enriched man materially, it has impoverished him humanly. Notwithstanding all propaganda and slogans about the Western world's faith in God, its idealism, its spiritual concern, our system has created a materialistic culture and a materialistic man. During his working hours, the individual is managed as part of a production team. During his hours of leisure time, he is managed and manipulated to be the perfect consumer who likes what he is told to like and yet has the illusion that he follows his own tastes. All the time he is hammered at by slogans, by suggestions, by voices of unreality which deprive him of the last bit of realism he may still have. From childhood on, true convictions are discouraged. There is little critical thought, there is little real feeling, and hence only conformity with the rest can save the individual from an unbearable feeling of loneliness and lostness. The individual does not experience himself as the active bearer of his own powers and inner richness, but as an impoverished "thing," dependent on powers outside of himself into which he has projected his living substance. Man is alienated from himself and bows down before the works of his own hands. He bows down before the things he produces, before the State and before the leaders of his own making. His own act becomes to him an alien power, standing over and against him instead of being ruled by him. More than ever in history the consolidation of our own product to an objective force above us, outgrowing our control, defeating our expectations, annihilating our calculations, is one of the main factors determining our development. His products, his machines, and the State have become the idols of modern man, and these idols represent his own life forces in alienated form.

Indeed, Marx was right in recognizing that "the place of all physical and mental senses has been taken by the self-alienation of all these senses, by the sense of having. Private property has made us so stupid and impotent that things become ours only if we *have* them, that is, if they exist for us as capital, and are owned by us, eaten by us, drunk by us; that is, used by us. We are poor in spite of all our wealth because we *have* much, but we *are* little."

As a result, the average man feels insecure, lonely, depressed, and suffers from a lack of joy in the midst of plenty. Life does not make sense to him; he is dimly aware that the meaning of life cannot lie in being nothing but a "consumer." He could not stand the joylessness and meaninglessness of life were it not for the fact that the system offers him innumerable avenues of escape, ranging from television to tranquilizers, which permit him to forget that he is losing more and more of all that is valuable in life. In spite of all slogans to the contrary, we are quickly approaching a society governed by bureaucrats who administer a mass-man, well fed, well taken care of, dehumanized and depressed. We produce machines that are like men and men who are like machines. That which was the greatest criticism of socialism fifty years ago—that it would lead to uniformity, bureaucratization, centralization, and a soulless materialism—is a reality of today's capitalism. We talk of freedom and democracy, yet an increasing number of people are afraid of the responsibility of freedom, and prefer the slavery of the well-fed robot; they have no faith in democracy and are happy to leave it to the political experts to make the decisions.

We have created a widespread system of communication by means of radio, television and newspapers. Yet people are misinformed and indoctrinated rather than informed about political and social reality. In fact, there is a degree of uniformity in our opinions and ideas which could be explained without difficulty if it were the result of political pressure and caused by fear. The fact is that all agree "voluntarily," in spite of the fact that our system rests exactly on the idea of the right to disagreement and on the predilection for diversity of ideas.

Doubletalk has become the rule in the free-enterprise countries, as well as among their opponents. The latter call dictatorship "people's democracies," the former call dictatorships "freedom-loving people" if they are political allies. Of the possibility of fifty million Americans being killed in an atomic attack, one speaks of the "hazards of war," and one talks of victory in a "showdown," when sane thinking makes it clear that there can be no victory for anyone in an atomic holocaust.

Education, from primary to higher education, has reached a peak. Yet, while people get more education, they have less reason, judgment, and conviction. At best their intelligence is improved, but their reason—that is, their capacity to penetrate through the surface and to understand the underlying forces in individual and social life—is impoverished more and more. Thinking is increasingly split from feeling, and the very fact that people tolerate the threat of an atomic war hovering over all mankind, shows that modern man has come to a point where his sanity must be questioned.

Man, instead of being the master of the machines he has built, has become their servant. But man is not made to be a thing, and with all the satisfactions of consumption, the life forces in man cannot be held in abeyance continuously. We have only one choice, and that is mastering the machine again, making production into a means and not an end, using it for the unfolding of man—or else the suppressed life energies will manifest themselves in chaotic and destructive forms. Man will want to destroy life rather than die of boredom.

Can we make our mode of social and economic organization responsible for this state of man? As was indicated above, our industrial system, its way of production and consumption, the relations between human beings which it fosters, creates precisely the human situation which has been described. Not because it *wants* to create it, not due to evil intentions of individuals, but because of the fact that the average man's character is formed by the practice of life which is provided by the structure of society.

No doubt the form which capitalism has taken in the twentieth century is very different from what it was in the nineteenth century—so different, in fact, that it is doubtful whether even the same term should be applied to both systems. The enormous concentration of capital in giant enterprises, the increasing separation of management from ownership, the existence of powerful trade unions, state subsidies for agriculture and for some parts of industry, the elements of the "welfare state," elements of price control and a directed market, and many more features radically distinguish twentieth-century capitalism from that of the past. Yet whatever terminology we choose, certain basic elements are common to the old and the new capitalism: the principle that not solidarity and love, but individualistic, egotistical action brings the best results for everybody; the belief that an impersonal mechanism, the market, should regulate the life of society, not the will, vision and planning of the people. Capitalism puts things (capital) higher than life (labor). Power follows from possession, not from activity. Contemporary capitalism creates additional obstacles for the unfolding of man. It needs smoothly working teams of workers, clerks, engineers, consumers; it needs them because big enterprises, led by bureaucracies, require this kind of organization and the "organization man" who fits into it. Our system must create people who fit its needs; it must create people who cooperate smoothly and in large numbers; people who want to consume more and more; people whose tastes are standardized and can be easily anticipated and influenced. It needs people who feel free and independent, not subject to any authority or principle of conscience, yet who are willing to be commanded to do what is expected of them, to fit into the social machine without friction; it needs people who can be guided without force, led without leaders, prompted without aim—except the aim to make good, to be on the move, to go ahead. Production is guided by the principle that capital investment must bring profit, rather than by the principle that the real needs of people determine what is to be produced. Since everything, including radio, television, books and medicines, is subject to the profit principle, the people are manipulated into the kind of consumption which is often poisonous for the spirit, and sometimes also for the body.

The failure of our society to fulfill the human aspirations rooted in our spiritual traditions has immediate consequences for the two most burning practical issues of our time: that of peace and that of the equalization between the wealth of the West and the poverty of two-thirds of mankind.

The alienation of modern man with all its consequences makes it difficult for him to solve these problems. Because of the fact that he worships things and has lost the reverence for life, his own and that of his fellow men, he is blind not only to moral principles, but also to rational thought in the interest of his survival. It is clear that atomic armament is likely to lead to universal destruction and, even if atomic war could be prevented, that it will lead to a climate of fear, suspicion, and regimentation which is exactly the climate in which freedom and democracy cannot live. It is clear that the economic gap between poor and rich nations will lead to violent explosions and dictatorships—yet nothing but the most half-hearted and hence futile attempts are suggested to solve these problems. Indeed, it seems that we are going to prove that the gods blind those whom they want to destroy.

Thus far goes the record of capitalism. What is the record of socialism? What did it intend and what did it achieve in those countries in which it had a chance of being realized?

Socialism in the nineteenth century, in the Marxian form and in its many other forms, wanted to create the material basis for a dignified human existence for everybody. It wanted work to direct capital, rather than capital to direct work. For socialism, work and capital were not just two economic categories, but rather they represented two principles: capital, the principle of amassed things, of *having;* and work, that of life and of man's powers, of *being* and becoming. Socialists found that in capitalism things direct life; that *having* is superior to *being;* that the past directs the present—and they wanted to reverse this relation. The aim of socialism was man's emancipation, his restoration to the unalienated, uncrippled individual who enters into a new, rich, spontaneous relationship with his fellow man and with nature. The aim of socialism was that man should throw away the chains which bind him, the fictions and unrealities, and transform himself into a being who can make creative use of his powers of feeling and of thinking. Socialism wanted man to become independent, that is, to stand on his own feet; and it believed that man can stand on his feet only if, as Marx said, "he owes his existence to himself, if he affirms his individuality as a total man in each of his relations to the world, seeing, hearing, smelling, tasting, feeling, thinking, willing, loving—in short, if he affirms and expresses all organs of his individuality." The aim of socialism was the union between man and man, and between man and nature.

Quite in contrast to the frequently uttered cliché that Marx and other socialists taught that the desire for maximal material gain was the most fundamental human drive, these socialists believed that it is the very structure of capitalist society which makes material interest the deepest motive, and that socialism would permit nonmaterial motives to assert themselves and free man from his servitude to material interests. (It is a sad commentary on man's capacity for inconsistency that people condemn socialism for its alleged "materialism," and at the same time criticize it with the argument that only the "profit motive" can motivate man to do his best.)

The aim of socialism was individuality, not uniformity; liberation from economic bonds, not making material aims the main concern of life; the experience of full solidarity of all men, not the manipulation and domination of one man by another. The principle of socialism was that each man is an end in himself and must never be the means of another man. Socialists wanted to create a society in which each citizen actively and responsibly participated in all decisions, and in which a citizen could participate because he was a person and not a thing, because he had convictions and not synthetic opinions.

For socialism not only is poverty a vice, but also wealth. Material poverty deprives man of the basis for a humanly rich life. Material wealth, like power, corrupts man. It destroys the sense of proportion and of the limitations which are inherent in human existence; it creates an unrealistic and almost crazy sense of the "uniqueness" of an individual, making him feel that he is not subject to the same basic conditions of existence as his fellow men. Socialism wants material comfort to be used for the true aims of living; it rejects individual wealth as a danger to society as well as to the individual. In fact, its opposition to capitalism is related to this very principle. By its very logic, capitalism aims at an ever-increasing material wealth, while socialism aims at an ever-increasing human productivity, aliveness, and happiness, and at material comfort only to the extent to which it is conducive to its human aims.

Socialism hoped for the eventual abolition of the state so that only things, and not people, would be administered. It aimed at a classless society in which freedom and initiative would be restored to the individual. Socialism, in the nineteenth century and until the beginning of the First World War, was the most significant humanistic and spiritual movement in Europe and America.

What happened to socialism?

It succumbed to the spirit of capitalism which it had wanted to replace. Instead of understanding it as a movement for the liberation of man, many of its adherents and its enemies alike understood it as being exclusively a movement for the economic improvement of the working class. The humanistic aims of socialism were forgotten, or only paid lip service to, while, as in capitalism, all the emphasis was laid on the aims of economic gain.

Just as the ideals of democracy lost their spiritual roots, the idea of socialism lost its deepest root—the prophetic-messianic faith in peace, justice, and the brotherhood of man.

Thus socialism became the vehicle for the workers to attain their place *within* the capitalistic structure rather than transcending it; instead of changing capitalism, socialism was absorbed by its spirit. The failure of the socialist movement became complete when in 1914 its leaders renounced international solidarity and chose the economic and military interests of their respective countries as against the ideas of internationalism and peace which had been their program.

The misinterpretation of socialism as a purely economic movement, and of nationalization of the means of production as its principal aim, occurred both in the right wing and in the left wing of the socialist movement. The reformist leaders of the socialist movement in Europe considered it their primary aim to elevate the economic status of the worker within the capitalist system, and they considered as their most radical measures the nationalization of certain big industries. Only recently have many realized that the nationalization of an enterprise is in itself not the realization of socialism, that to be managed by a publicly appointed bureaucracy is not basically different for the worker than being managed by a privately appointed bureaucracy.

The leaders of the Communist Party in the Soviet Union interpreted socialism in the same purely economic way. But living in a country much less developed than western Europe and without a democratic tradition, they applied terror and dictatorship to enforce the rapid accumulation of capital, which in western Europe had occurred in the nineteenth century. They developed a new form of state capitalism which proved to be economically successful and humanly destructive. They built a bureaucratically managed society in which class distinction—both in an economic sense and as far as the power to command others is concerned—is deeper and more rigid than in any of the capitalist societies of today. They define their system as socialistic because they have nationalized the whole economy, while in reality their system is the complete negation of all that socialism stands for—the affirmation of individuality and the full development of man. In order to win the support of the masses who had to make unendurable sacrifices for the sake of the fast accumulation of capital, they used socialistic, combined with nationalistic, ideologies and this gained them the grudging cooperation of the governed.

Thus far the free-enterprise system is superior to

the communist system because it has preserved one of the greatest achievements of modern man—political freedom—and with it a respect for the dignity and individuality of man, which links us with the fundamental spiritual tradition of humanism. It permits possibilities of criticism and of making proposals for constructive social change which are practically impossible in the Soviet police state. It is to be expected, however, that once the Soviet countries have achieved the same level of economic development that western Europe and the United States have achieved—that is, once they can satisfy the demand for a comfortable life—they will not need terror, but will be able to use the same means of manipulation which are used in the West: suggestion and persuasion. This development will bring about the convergence of twentieth-century capitalism and twentieth-century communism. Both systems are based on industrialization; their goal is ever-increasing economic efficiency and wealth. They are societies run by a managerial class and by professional politicians. They are both thoroughly materialistic in their outlook, regardless of lip service to Christian ideology in the West and secular Messianism in the East. They organize the masses in a centralized system, in large factories, in political mass parties. In both systems, if they go on in the same way, the mass-man, the alienated man—a well-fed, well-clothed, well-entertained automaton-man governed by bureaucrats who have as little a goal as the mass-man has—will replace the creative, thinking, feeling man. *Things* will have the first place, and man will be dead; he will *talk* of freedom and individuality, while he will *be* nothing.

Where do we stand today?

Capitalism and a vulgarized, distorted socialism have brought man to the point where he is in danger of becoming a dehumanized automaton; he is losing his sanity and stands at the point of total self-destruction. Only full awareness of his situation and its dangers and a new vision of a life which can realize the aims of human freedom, dignity, creativity, reason, justice, and solidarity can save us from almost certain decay, loss of freedom, or destruction. We are not forced to choose between a managerial free-enterprise system and a managerial communist system. There is a third solution, that of democratic, humanistic socialism which, based on the original principles of socialism, offers the vision of a new, truly human society.

IV.

Humanist Socialism

On the basis of the general analysis of capitalism, communism, and humanistic socialism, a socialist program must differentiate between three aspects: What are the *principles* underlying the idea of a socialist party? What are the *intermediate goals* of humanistic socialism for the realization of which socialists work? What are the immediate *short-range goals* for which socialists work, as intermediate goals have not yet been achieved?

What are the *principles* which underlie the idea of a humanistic socialism? Every social and economic system is not only a specific system of relations *between things and institutions*, but a system of *human relations*. Any concept and practice of socialism must be examined in terms of the kind of relations between human beings to which it is conducive.

The supreme value in all social and economic arrangements is man; the goal of society is to offer the conditions for the full development of man's potentialities, his reason, his love, his creativity; all social arrangements must be conducive to overcoming the alienation and crippledness of man, and to enable him to achieve real freedom and individuality. The aim of socialism is an association in which the full development of each is the condition for the full development of all.

The supreme principle of socialism is that man takes precedence over things, life over property, and hence, work over capital; that power follows creation, and not possession; that man must not be governed by circumstances, but circumstances must be governed by man.

In relations between people, every man is an end in himself and must never be made into a means to another man's ends. From this principle it follows that nobody must personally be subject to anyone because he owns capital.

Humanistic socialism is rooted in the conviction of the unity of mankind and the solidarity of all men. It fights any kind of worship of State, nation, or class. The supreme loyalty of man must be to the human race and to the moral principles of humanism. It strives for the revitalization of those ideas and values upon which Western civilization was built.

Humanistic socialism is radically opposed to war and violence in all and any forms. It considers any attempt to solve political and social problems by force and violence not only as futile, but as immoral and inhuman. Hence it is uncompromisingly opposed to any policy which tries to achieve security by armament. It considers peace to be not only the absence of war, but a positive principle of human relations based on free cooperation of all men for the common good.

From socialist principles it follows not only that each member of society feels responsible for his fellow citizens, but for all citizens of the world. The injustice which lets two-thirds of the human race live in abysmal poverty must be removed by an effort far beyond the ones hitherto made by wealthy nations to help the underdeveloped nations to arrive at a humanly satisfactory economic level.

Humanistic socialism stands for freedom. It stands for freedom from fear, want, oppression, and violence. But freedom is not only *from*, but also freedom *to*; freedom to participate actively and responsibly in all decisions concerning the citizen, freedom to develop the individual's human potential to the fullest possible degree.

Production and consumption must be subordinated to the needs of man's development, not the reverse. As a consequence all production must be directed by the principle of its social usefulness, and not by that of its material profit for some individuals or corporations. Hence if a choice has to be made between greater production on the one hand, and greater freedom and human growth on the other, the human as against the material value must be chosen.

In socialist industrialism the goal is not to achieve the highest *economic* productivity, but to achieve the highest *human* productivity. This means that the way in which man spends most of his energy, in work as well as in leisure, must be meaningful and interesting to him. It must stimulate and help to develop *all* his human powers—his intellectual as well as his emotional and artistic ones.

While, in order to live humanly, basic material needs must be satisfied, consumption must not be an aim in itself. All attempts to stimulate material needs artificially for the sake of profit must be prevented. Waste of material resources and senseless consumption for consumption's sake are destructive to mature human development.

Humanistic socialism is a system in which man governs capital, not capital man; in which, so far as it is possible, man governs his circumstances, not circumstances man;

in which the members of society plan what they want to produce, rather than have their production follow the laws of the impersonal powers of the market and of capital with its inherent need for maximum profit.

Humanistic socialism is the extension of the democratic process beyond the purely political realm, into the economic sphere; it is political *and* industrial democracy. It is the restoration of political democracy to its original meaning: the true participation of informed citizens in all decisions affecting them.

Extension of democracy into the economic sphere means democratic control of all economic activities by the participants: manual workers, engineers, administrators, etc. Humanistic socialism is not primarily concerned with legal ownership, but with social control of the large and powerful industries. Irresponsible control by bureaucratic management representing the profit interest of capital must be replaced by administration acting on behalf of, and controlled by, those who produce and consume.

The aim of humanistic socialism can be attained only by the introduction of a maximum of decentralization compatible with a minimum of centralization necessary for the coordinated functioning of an industrial society. The functions of a centralized state must be reduced to a minimum, while the voluntary activity of freely cooperating citizens constitutes the central mechanism of social life.

While the basic general aims of humanistic socialism are the same for all countries, each country must formulate its own specific aims in terms of its own traditional and present situation, and devise its own methods to achieve this aim. The mutual solidarity of socialist countries must exclude any attempt on the part of one country to impose its methods on another. In the same spirit, the writings of the fathers of socialist ideas must not be transformed into sacred scriptures which are used by some to wield authority over others; the spirit common to them, however, must remain alive in the hearts of socialists and guide their thinking.

Humanistic socialism is the voluntary, logical outcome of the operation of human nature under rational conditions. It is the realization of democracy, which has its roots in the humanistic tradition of mankind, under the conditions of an industrial society. It is a social system which operates without force, neither physical force nor that of hypnoid suggestions by which men are forced without being aware of it. It can be achieved only by appealing to man's reason, and to his longing for a more human, meaningful, and rich life. It is based on faith in man's ability to build a world which is truly human, in which the enrichment of life and the unfolding of the individual are the prime objects of society, while economics is reduced to its proper role as the means to a humanly richer life.

In discussing the goals of humanistic socialism we must differentiate between the *final* socialist goal of a society based on the free cooperation of its citizens and the reduction of centralized State activity to a minimum, and the *intermediate* socialist goals before this final aim is reached. The transition from the present centralized State to a completely decentralized form of society cannot be made without a transitory period in which a certain amount of central planning and State intervention will be indispensable. But in order to avoid the dangers that central planning and State intervention may lead to, such as increased bureaucratization and weakening of individual integrity and initiative, it is necessary: a) that the State is brought under the efficient control of its citizens; b) that the social and political power of the big corporations is broken; c) that from the very beginning all forms of decentralized, voluntary associations in production, trade, and local social and cultural activities are promoted.

While it is not possible today to make concrete and detailed plans for the final socialist goals, it is possible to formulate in a tentative fashion the intermediate goals for a socialist society. But even as far as these intermediate goals are concerned, it will take many years of study and experimentation to arrive at more definite and specific formulations, studies to which the best brains and hearts of the nation must be devoted.

Following the principle that social control and not legal ownership is the essential principle of socialism, its first goal is the transformation of all big enterprises in such a way that their administrators are appointed and fully controlled by all participants—workers, clerks, engineers—with the participation of trade union and consumer representatives. These groups constitute the highest authority for every big enterprise. They decide all basic questions of production, price, utilization of profits, etc. The stockholders continue to receive an appropriate compensation for the use of their capital, but have no right of control and administration.

The autonomy of an enterprise is restricted by central planning to the extent to which it is necessary to make production serve its social ends.

Small enterprises should work on a cooperative basis, and they are to be encouraged by taxation and other means. Inasmuch as they do not work on a cooperative basis, the participants must share in the profits and control the administration on an equal basis with the owner.

Certain industries which are of basic importance for the whole of society, such as oil, banking, television, radio, medical drugs, and transportation, must be nationalized; but the administration of these nationalized industries must follow the same principles of effective control by participants, unions and consumers.

In all fields in which there is a social need but not an adequate existing production, society must finance enterprises which serve these needs.

The individual must be protected from fear and the need to submit to anyone's coercion. In order to accomplish this aim, society must provide, free for everyone, the minimum necessities of material existence in food, housing, and clothing. Anyone who has higher aspirations for material comforts will have to work for them, but the minimal necessities of life being guaranteed, no person can have power over anyone on the basis of direct or indirect material coercion.

Socialism does not do away with individual property for use. Neither does it require the complete leveling of income; income should be related to effort and skill. But differences in income should not create such different forms of material life that the life experience of one cannot be shared by, and thus remains alien to, another.

The principle of political democracy must be implemented in terms of the twentieth-century reality. Considering our technical instrumentalities of communication and tabulation, it is possible to reintroduce the principle of the town meeting into contemporary mass society. The forms in which this can be accomplished need study and experimentation. They may consist of the formation of hundreds of thousands of small face-to-face groups (organized along the principle of place of work or place of residence) which would constitute a new type of Lower House, sharing decision-making with a centrally elected parliament. Decentralization must strive at putting important decisions into the hands of the inhabitants of small, local areas which are still subject to the fundamental principles which govern the life of the whole society. But whichever forms are to be found, the essential principle is that the democratic process is transformed into one in which well-informed and responsible citizens—not automatized mass-men, controlled by the methods of hypnoid mass suggestion—express their will.

Not only in the sphere of political decisions, but with regard to all decisions and arrangements, the grip of the bureaucracy must be broken in order to restore freedom. Aside from decisions which filter down from above, activity in all spheres of life on the grass-roots level must be developed which can "filter up" from below to the top. Workers organized in unions, consumers organized in consumers' organizations, citizens organized in the above-mentioned face-to-face political units, must be in constant interchange with central authorities. This interchange must be such that new measures, laws, and provisions can be suggested and, after voting, decided from the grass roots, and that all elected representatives are subject to continuous critical appraisal and, if necessary, recall.

According to its basic principles, the aim of socialism is the abolition of national sovereignty, the abolition of any kind of armed forces, and the establishment of a commonwealth of nations.

In the sphere of education, the main aims are those of helping to develop the critical powers of the individual and to provide a basis for the creative expression of his personality—in other words, to nurture free men who will be immune to manipulation and to the exploitation of their suggestibility for the pleasure and profit of others. Knowledge should not be a mere mass of information, but the rational means of understanding the underlying forces that determine material and human processes. Education should embrace not only reason but the arts. Capitalism, as it has produced alienation, has divorced and debased both man's scientific understanding and his aesthetic perception. The aim of socialist education is to restore man to the full and free exercise of both. It seeks to make man not only an intelligent spectator but a well-equipped participant, not only in the production of material goods, but in the enjoyment of life. To offset the dangers of alienated intellectualization, factual and theoretical instruction shall be supplemented by training in manual work and in the creative arts, combining the two in craftsmanship (the production of useful objects of art), in primary and secondary education. Each adolescent must have had the experience of producing something valuable with his own hands and skills.

The principle of irrational authority based on power and exploitation must be replaced, not by a laissez-faire attitude, but by an authority which is based on the competence of knowledge and skill—not on intimidation, force, or suggestion. Socialist education must arrive at a new concept of rational authority which differs both from irrational authoritarianism and from an unprincipled laissez-faire attitude.

Education must not be restricted to childhood and adolescence, but the existing forms of adult education must be greatly enlarged. It is especially important to give each person the possibility of changing his occupation or profession at any time of life; this will be economically possible if at least his minimal material needs are taken care of by society.

Cultural activities must not be restricted to providing intellectual education. All forms of artistic expression (through music, dance, drama, painting, sculpture, architecture, etc.) are of paramount importance for the human development of man. Society must channel considerable means for the creation of a vast program of artistic activities and useful as well as beautiful building programs, even at the expense of other and less important consumer satisfactions. Great care should be taken, however, to conserve the integrity of the creative artist, to avoid turning socially responsible art into bureaucratic or "State" art. A healthy balance must be maintained between the legitimate claims of the artist upon society and its legitimate claims upon him. Socialism seeks to narrow the gap between the producer and the consumer in the realm of art and seeks ultimately to eliminate this distinction so far as possible by creating optimum conditions for the flourishing of every individual's creative potentialities. But it holds up no preconceived pattern and recognizes that this is a problem that will require much more study than has been given to it up till now.

Complete equality of races and sexes is a matter of course for a socialist society. This equality, however, does not imply sameness, and every effort must be made to permit the fullest development of the gifts and talents peculiar to each racial and national group, as well as to the two sexes.

Freedom of religious activities must be guaranteed, together with the complete separation of State and Church.

The foregoing program is meant to serve as a guide to the principles and goals of socialism. Its concrete and detailed formulation requires a great deal of discussion. To conduct this discussion and to arrive at concrete and detailed suggestions is one of the main tasks of a socialist party. Such discussion requires examining all data which practical experience and the social sciences can bring forward. But first of all it requires imagination and the courage to see new possibilities, instead of the outworn routine of thinking.

Quite aside from this, it will take considerable time until the majority of the people in the United States will be convinced of the validity of socialist principles and goals. What is the task and function of a socialist party during the time before it has succeeded in this task?

The SP-SDF (Socialist Party-Social Democratic Federation)* must embody in its own structure and activities the very principles it stands for; it must not only strive for the achievement of socialism in the future, but must begin with its realization in its own midst immediately. Hence the SP-SDF must not try to convince the people of its program by appeal to irrational emotions, hypnoid suggestions, or "attractive personalities," but by the realism, correctness, and penetration of its analysis of economic, social, political, and human situations. The SP-SDF must become the moral and intellectual conscience of the United States and divulge its analyses and judgments in the widest possible manner.

The conduct of activities of the SP-SDF must follow its principles in the sense of the optimum of decentralization and the active, responsible participation of its members in discussions and decisions. It must also give full scope to the expression and divulgence of minority opinions. The socialist program cannot be a fixed plan, but must grow and develop through the continuous activity, effort, and concern of the members of the party.

The SP-SDF thus must be different from other political parties, not only in its program and ideals, but in its

* Fromm's considerations of the psychological dimensions of social and political phenomena led him to support for some time this American Socialist party.

very structure and way of functioning. It must become a spiritual and social home for all its members who are united in the spirit of humanistic realism and sanity, and by the solidarity of the common concern for and the common faith in man and his future.

The SP-SDF must develop an extensive educational campaign among workers, students, professionals, and members of all social classes who can be expected to have a potential understanding for socialist criticism and socialist ideals.

The SP-SDF cannot expect to gain victory in a short time. But this does not mean that it should not aim at the widest social influence and power. It must strive to gain the allegiance of an ever-increasing number of people who, through the party, make their voices heard within the United States and throughout the whole world.

The SP-SDF is rooted in the humanistic tradition of socialism; it strives for the transformation of the traditional socialist goals to fit the conditions of twentieth-century society as a condition for their realization. Particularly it rejects the ideas of achieving its goals by force or by the establishment of any kind of dictatorship. Its only weapons are the realism of its ideas, the fact that they appeal to the true needs of man, and the enthusiastic allegiance which those citizens will give it who have seen through the fictions and delusions which fill the minds of people today, and who have faith in a richer, fuller life.

It is not enough that the members of the SP-SDF believe in a common ideal. Such faith becomes empty and sterile if it is not translated into action. The life of the party must be organized in such a manner that it offers ample and varied possibility for every member to translate his concern into meaningful and immediate action. How can this be done?

It must be understood clearly that the basic goals of socialism, especially the method of management of big enterprises by the participants, union and consumers' representatives, the revitalization of the democratic process, the guaranteed minimum for existence for every citizen, constitute problems the details of which are exceedingly difficult to solve. Their solution requires basic theoretical research in the fields of economics, work organization, psychology, etc.; and, in addition, it requires practical plans and experimentation. If these social problems are approached in the same spirit of faith and imagination which exists among the natural scientists and technicians, solutions will be found which, looked at from the present situation, might appear as fantastic as space travel appeared twenty years ago. Yet the difficulties of arriving at a solution for a sane and human social organization are not any greater than those in the fields of the theoretical and applied natural sciences.

The first task, then, for socialists is to study the problems of applied socialism within their own sphere of activity and to discuss their experiences and suggestions for socialist solutions in the working units of their SP-SDF. Supplementing this group activity are permanent committees for the investigation of these problems. These committees will be composed of specialists in the various fields of economics, sociology, psychology, foreign policy, etc. The committees of investigation and the working units will be in close mutual contact, exchanging their ideas and experiences, and thus stimulating each other.

But the activities of the members of the SP-SDF must not be restricted to imaginative thinking and planning. Beyond this, immediate and concrete action is necessary. It is important that each member demonstrates the socialist way of life in his or her place of work, wherever it may be—in factories, offices, schools, laboratories, hospitals, etc. Each member must demonstrate the socialist way of approaching problems by his own way of dealing with them and by stimulating others. It is especially important that the members of the SP-SDF who are union members work actively for greater member activity and participation in the life of the trade unions. Inside and outside the trade unions, the members of the SP-SDF will support all tendencies for decentralization, active grass-roots participation, and fight all forms of bureaucratism.

The SP-SDF wants to attract men and women who are genuinely concerned with the problem of the humanization of society and who, out of this concern, work for it and are willing to make the sacrifice in time and money which this work requires.

Although the SP-SDF has its center in the fundamental goals of its programs, it will participate actively in the furtherance of all immediate political aims which are of importance for the progressive development of our society. It will cooperate with all political groupings and individuals that sincerely strive for the same aims. Among these aims are, in particular:

- A sane foreign policy, based on a realistic appreciation of the given facts of political life—a policy which seeks reasonable compromise and realizes that war can be averted only if the two power blocs accept their present economic and political positions and renounce every attempt to change them by force.

- Fight against the idea that our security can be gained by armaments. The only way to avoid total destruction lies in total disarmament. This implies that disarmament negotiations must not be used to prevent real disarmament, but that we must be willing to take risks in the attempt to achieve it.

- A program of economic aid to underdeveloped countries on an immensely larger scale than our present one, and at the cost of considerable sacrifice on the part of our citizens for the accomplishment of such a program. We advocate a policy which does not serve the interests of American capital investments in foreign countries and does not involve United States foreign policy in indirect interference with the independence of small nations.

- Strengthening of the United Nations and of all efforts to use its assistance in the solving of international disputes and in large-scale foreign aid.

- Support of all measures to raise the standard of living of that part of our population which is still living below the material standard of the majority. This applies to poverty caused by economic as well as by regional and racial factors.

- Support of all efforts for decentralization and grassroots activities. This implies support of all attempts to curb irresponsible power in corporate, governmental, and union bureaucracy.

- Support of all measures for social security which lead to immediate relief in distress situations caused by unemployment, sickness, and old age. Support of all measures in the direction of socialized medicine, with the understanding that the free choice of doctors and a high level of medical services must be upheld.

- Economic measures which lead to the full use of our agricultural productive capacity and our surplus, nationally and internationally.

- Support of measures to set up an economic commission consisting of representatives of industry, commerce, trade unions, economists, and consumer representatives. This commission should be charged with undertaking a regular examination of the needs of our economy and developing overall plans for changes in the interest of the nation as a whole. Its most immediate task would be to discuss and propose plans for the change from armaments to peace production. The reports of this commission, including minority reports, should be published and distributed widely. Similar commissions should be convoked in the field of foreign policy, culture, and education; the members of these commissions should represent wide sectors of the population, and consist of men whose knowledge and integrity are generally recognized.

- Vast governmental expenditures for housing, road building, and hospital construction, and for cultural activities such as music, theater, dance, and art.

- Given the wealth of the United States, we can begin to experiment socially. State-owned enterprises must be organized in which various forms of workers' participation in management are tried out.

- In industries of basic social importance, the government must organize yardstick enterprises, which compete with private industry and in this way force it to raise its standards. This must be done first of all in the field of radio, television and movies, and in other fields if desirable.

- Efforts must be made to begin with a program of workers' participation in the management of the big corporations. Twenty-five percent of the votes on the decision-making boards should be given to workers and employees, freely elected in each enterprise.

- The influence of the unions must be strengthened, not only with regard to the problems of wages, but also with regard to their influence on problems of working conditions, etc. Simultaneously a process of democratization within the unions must be furthered with all energy.

- All attempts must be supported which aim at the restriction of hypnoid suggestion in commercial and political propaganda.

We are aware that the above-mentioned program refers mainly to industrialized countries like those of North America and Europe. For all other countries the program must vary according to their specific conditions. However, the general principles underlying this program, that of production for social use, the strengthening of an effective democratic process, industrially as well as politically, are valid for all countries.

We appeal to every citizen to feel his responsibility for his life, that of his children and that of the whole human family. Man is on the verge of the most crucial choice he has ever made: whether to use his skill and brain to create a world which can be, if not a paradise, at least a place for the fullest realization of man's potentialities, a world of joy and creativity—or a world which will destroy itself either with atomic bombs or through boredom and emptiness.

Indeed, socialism differs from other party programs in that it has a vision, an ideal for a better, more human society than the present one. Socialism does not want only to improve this or that defect of capitalism, it wants to accomplish something which does not yet exist; it aims at a goal which transcends the given empirical social reality, yet which is based on a real potentiality. Socialists have a vision and say: this is what we want; this is what we strive for; it is not the absolute and the final form of life, but it is a much better, more human form of life. It is the realization of the ideals of humanism which have inspired the greatest achievements of Western and Eastern culture.

Many will say that people do not want ideals, that they do not want to go beyond the frame of reference in which they live. We socialists say that this is not true.

On the contrary, people have a deep longing for something they can work for and have faith in. Man's whole vitality depends on the fact that he transcends the routine part of his existence, that he strives for the fulfillment of a vision which is not impossible to realize—even though it has not yet been achieved. If he has no chance to strive for a rational, humanistic vision, he will eventually, worn out and depressed by the boredom of his life, fall prey to the irrational satanic visions of dictators and demagogues. It is exactly the weakness of contemporary society that it offers no ideals, that it demands no faith, that it has no vision—except that of more of the same. We socialists are not ashamed to confess that we have a deep faith in man and in a vision of a new, human form of society. We appeal to the faith, hope and imagination of our fellow citizens to join us in this vision and in the attempt to realize it. Socialism is not only a socioeconomic and political program; it is a human program: *the realization of the ideals of humanism under the conditions of an industrial society.*

Socialism must be radical. To be radical is to go to the roots; and the root is Man.

About the Author

ERICH FROMM (1900–1980) emigrated from Germany in 1934 to the United States, where he held a private practice and taught at Columbia, Yale, and New York University. His many books include *The Art of Loving, Escape from Freedom, Man for Himself,* and *The Anatomy of Human Destructiveness.*

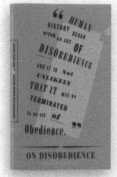